Nathan & Grandpa's Book of Space

By

Nathan Williams (age 6)

&

Grandpa

To Mummy and Daddy,

my lovely sister Emily,

Grandma Liz,

Granddad Robin and to the memory of my

Grandma Catherine.

Pictures drawn and coloured by Nathan.

Words spoken by Nathan to Grandpa.

I love space and I wanted to write this book to share everything I know.

The Sun

The sun is very hot. It is 50 million degrees Celsius.

The sun is our star.

The sun produces heat and light that enables life to grow on Earth.

The sun has big solar flares of molten plasma gas that are extremely hot!

The Moon

The Moon is made of rock.

It has lots of craters because meteors hit it a lot when the Moon was new.

The Moon was formed from lots of rocks and dust that came together when the Earth was formed.

The Earth

The Earth is the only planet where we know life can exist because it has water, heat and light.

The Earth's atmosphere has lots of oxygen.

The Earth has big seas and land in large continents.

Life on Earth includes flowers, humans, and hundreds and hundreds of different animals, insects, trees, bugs and bacteria.

The Big Bang

The Big Bang happened billions of years ago - way before Grandpa was born.

The whole universe exploded out of nowhere.

Nobody really knows how that happened.

It is a mystery.

Space

Space fills the universe.

There are billions of galaxies and trillions of stars in space.

Space is humungous!!

It has meteors, planets, nothing, dust, comets, black holes and the International Space Station.

Drawn by Grandpa – coloured by Nathan

Stephen Hawkins

Stephen Hawkins is my favourite scientist.

He knows everything about space.

Even though he is in a wheelchair and has to speak through a computer he is very clever.

He is cleverer than grandpa (only just!).

Stephen Hawkins tells us about the universe, space and black holes.

Mars

Mars is the red planet.

It has the biggest volcano in the solar system.

The scientists sent a mission to Mars with a robot called Rover which takes pictures and collects rock.

They think there might be life there because there used to be water in the dried up rivers and lakes.

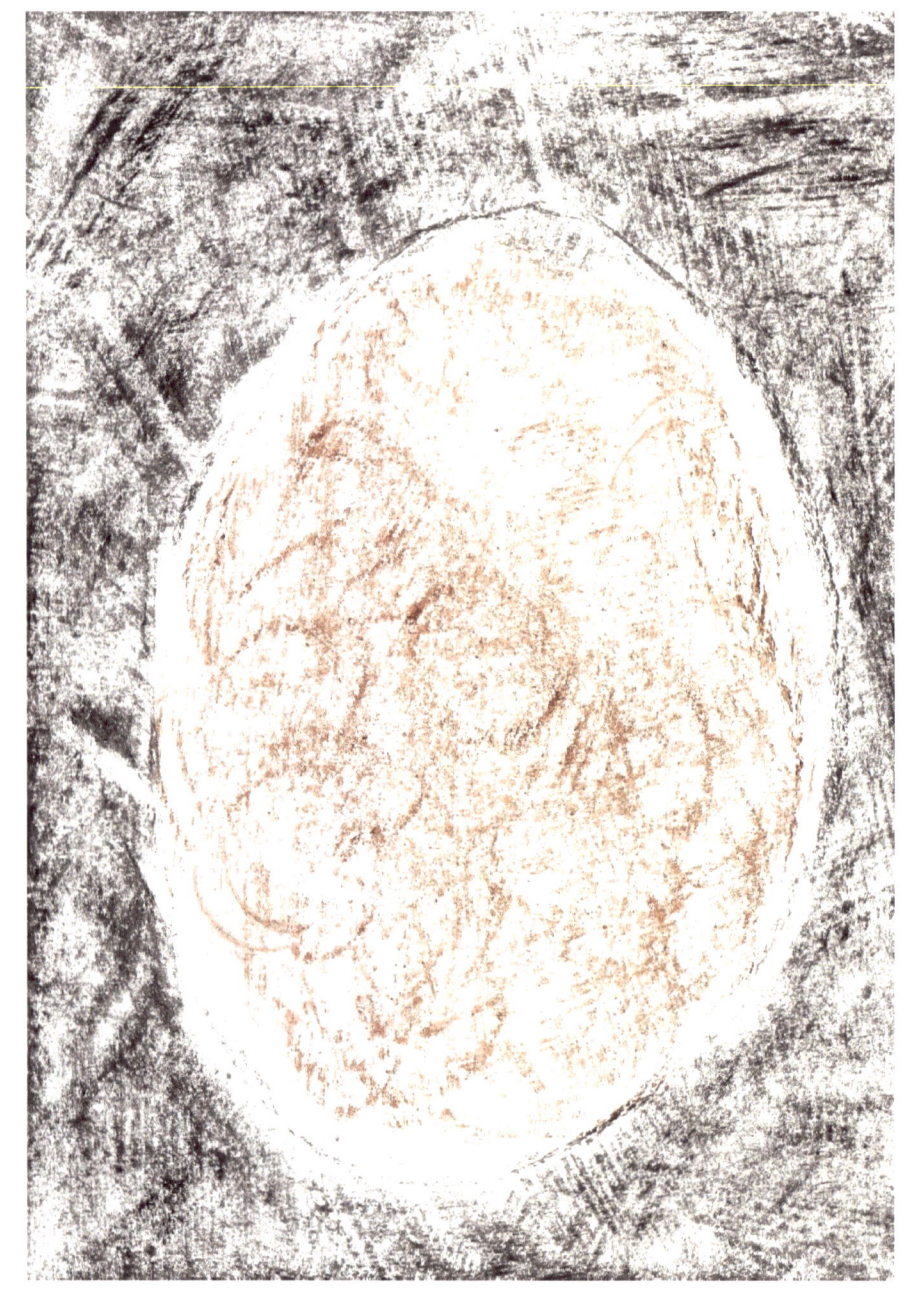

Venus

Venus is the planet that is second closest to the Sun.

Even though Venus is not the closest to the Sun it is the hottest planet because it has a huge atmosphere of carbon dioxide and water vapour.

The atmosphere keeps the heat in.

You cannot see Venus because of all the thick clouds!

Space Ships

Some space ships are launched from space shuttles. Some are rockets which blast off into space. These go the furthest because they have different parts that drop off when they have burnt out and fall back to Earth.

Apollo 11 was a space rocket.

The people in a rocket are in the nose-cone that goes all the way to the International Space Station.

Space ships have landed people on the Moon.

Space ships have been on missions to Mars and a comet.

Aliens

Nobody really knows if aliens are real or not.

That is why I like talking about aliens.

Aliens are creatures that live on other planets.

There might be clever aliens somewhere in the universe.

We might get to meet them when we have faster rockets.

Stephen Hawkins thought of a rocket that was fast enough to go to the stars.

Black Holes

Black Holes are made when huge stars die and collapse in on themselves.

They form a Black Hole that sucks things in with their gravity.

They suck in stars and they even suck in light!

That is gravity's dark side!

Light cannot escape from a Black Hole. It is stronger than Grandpa!

Meteors

Meteors are space rocks.

They are extremely fast – faster than a space rocket!

They crash into planets!

Sometimes they crash into the Earth and destroy things.

Meteors wiped out the dinosaurs.

Once there was a huge meteor that was heading straight for Earth. It was lucky that the Earth has a protector called Jupiter. Jupiter is a big planet, a gas giant that has big gravity. It pulled the meteor in and saved the Earth!

Saturn   Jupiter   Uranus   Neptune

Mercury

The rest of the planets

Our solar system has eight planets.

Four planets are made of rock. They are the rocky planets – Earth, Mercury, Venus and Mars.

The gas giants are Jupiter, Saturn, Uranus and Neptune.

There are dwarf planets as well but they are not really planets. One of them is called Pluto. Two others are called Ceres and Eris.

Stars

Stars are really very big. They are suns a long way away. They are usually bigger than our sun.

There are trillions and trillions and zillions of stars.

When you look up at the sky you don't just see white stars. If you look very carefully you can see red stars, blue stars and yellow stars.

Some of these stars have solar systems with planets.

They might even have aliens living on their planets!

Galaxies

Galaxies are made up of billions of stars and clouds of dust.

They swirl around in big spirals. But there are different types of galaxies.

Sometimes galaxies smash together and become bigger galaxies.

We see some galaxies from the side and they look flat.

Some of the stars we see are really galaxies of trillions of stars.

The Solar System

Our solar system is made up of eight planets that orbit the sun.

The closest planet to the sun is Mercury, then Venus, then Earth, then Mars, then Jupiter, then Saturn, Uranus and last of all Neptune.

There is an asteroid belt between Mars and Jupiter made up of space rocks.

Some planets have moons. Earth has one moon.

Sometimes meteors and comets crash into our solar system.

One day I will probably be a space scientist.

I may be able to go into space and find aliens!